# YOUR KNOWLEDGE HAS VALUE

- We will publish your bachelor's and
  master's thesis, essays and papers

- Your own eBook and book -
  sold worldwide in all relevant shops

- Earn money with each sale

Upload your text at www.GRIN.com
and publish for free

# Localisation and Identification of Ion Transport Peptide in the Brain of the Cockroach Leucophaea maderae

Janna Witt

**Bibliographic information published by the German National Library:**

The German National Library lists this publication in the National Bibliography; detailed bibliographic data are available on the Internet at http://dnb.dnb.de.

ISBN: 9783346806161
This book is also available as an ebook.

© GRIN Publishing GmbH
Nymphenburger Straße 86
80636 München

All rights reserved

Print and binding: Books on Demand GmbH, Norderstedt, Germany
Printed on acid-free paper from responsible sources.



GRIN web shop: https://www.grin.com/document/1322201

Berufskolleg Hilden des Kreises Mettmann

Biological Technical Assistants

Report

# Localisation and Identification of Ion Transport Peptide in the Brain of the Cockroach Leucophaea maderae

<table>
<tr><td>Trainee</td><td>Work experience placement</td></tr>
<tr><td>Janna Witt</td><td>Zoologiska Institutionen: Funktionell zoomorfologi<br>Stockholms universitet</td></tr>
</table>

This report has been submitted on:
30.10.2012

Work placement:
9th of July till 7th of October 2012

**Table of Contents**

## Abstract

The neurohormone Crustacean Hyperglycaemic Hormone (CHH) is the major member of the CHH-superfamily and controls the osmoregulation, moulting and reproduction in decapod crustaceans.

In some insects a neurohormone occurs, which is closely related to CHH; the Ion Transport Peptide (ITP).

In locusts it is a chloride transport-stimulating and acid secretion-inhibiting hormone and controls (together with other peptide hormones) the final modification of the primary urine by ion, solute and water reabsorptive processes primarily in the hindgut.

As ITP is well known in locusts, fruit flies and moths, we want to show its existance in another important model organism for the neuropeptide research: the cockroach Leucophaea maderae (L.maderae).

To localise ITP and the neurosecretory cells, which produce/release ITP, in the brain tissue of L.maderae, we use two indirect immunohistochemical techniques: immunofluorescence staining and peroxidase-antiperoxidase staining. The basic principle of these methods are predicated on the process of detecting antigens in cells of a tissue section by exploiting the principle of antibodies binding specifically to antigens in biological tissues. Because of the similarity to CHH, we can use an antiCHH antibody to stain ITP in the tissue.

To identify ITP in the brain of L.maderae, we conduct a reversed-phase high-performance liquid chromatography (HPLC) and a following direct enzyme-linked immunosorbent assay (ELISA) of 10 retrocerebral complexes.

Our work has shown that there is ITP in the brain of the cockroach L.maderae. We were not only able to demonstrate that ITP appears in the retrocerebral complexes (by HPLC and ELISA), but we also showed where ITP is located in the brain and the surrounding periphery (by the immunohistochemical methods).

We found a lot of CHH-immunoreactive structures and cells in the protocerebrum, superior lateral protocerebrum, retrocerebral complex, around the sub-oesophageal ganglion and muscle cells.

Also, our results maybe give some hints about the function of ITP in the cockroach L.maderae.

## 1. Introduction

Neurohormones are a group of substances produced by specialized nerve cells, the neurosecretory cells, that are structurally similar to nervous rather than endocrine system constituents. The neurohormones pass along axons and are released into the blood- or haemolymph stream at special regions called neurohemal organs. Neurohormones thus constitute a linkage between sensory stimuli and chemical responses (Encyclopaedia Britannica article: Neurohormone).

In crustaceans, the crustacean hyperglycaemic hormone (CHH) is a neurohormone and the major member of the CHH-superfamily type-I. It has several functions such as controlling osmoregulation, moulting and reproduction in decapod crustaceans (Webster et al, 2012). Closely related to CHH is the ion transport peptide (ITP), occurring in insects. CHH and ITP have many characteristic features in common: 1) three disulphide bonds, 2) normally a length of 72-73 amino acids, 3) C-terminal amidation, 4) presence of aromatic amino acids in postition 3 of the N-terminal putative a-helix and 5) same core structure of the first 40 or 41 amino acids (Dircksen, 2008, Webster et al, 2012).

ITP is one of the antidiuretic factors found first in locusts and later in other insects (Dircksen et al, 2008). In locusts, it is a chloride transport-stimulating and acid secretion-inhibiting hormone and controls (together with other peptide hormones) the final modification of the primary urine by ion, solute and water reabsorptive processes primarily in the hindgut. It is also discussed to have an ecdysis-related function in moths (Dircksen, 2008).

In locusts, moth and fruit flies there are two types of ITPs: ITP and the slightly longer isoform ITPL. Both arise by alternative splicing from single ITP genes, that are similar in several insects and crustaceans.

Four principal neuron types, which contain/release ITP/ITPL, are localisated in the same insects as mentioned above in 1) neurosecretory cells of the pars lateralis/retrocerebral complex (usually one splice form only), 2) interneurons (either one of the splice forms), 3) hindgut-innervating abdominal ITP neurons (Drosophila only) and 4) intrinsic, putative sensory neurosecretory cells in peripheral neurohaemal perisympathetic/perivisceral organs or transverse nerves most likely containing ITPLs (Dircksen, 2008).

As ITP is well known in locusts, fruit flies and moths, we wanted to show its existance in another important model organism for the neuropeptide research: the cockroach Leucophaea maderae (L.maderae).

Like all insects, L.maderae looses water through the soft cuticule, and it lives in areas where it is often exposed to water shortage. Therefore, L.maderae is dependent upon saving and/or regaining of water and needs a hormone such as ITP. The primary aim of this study is to localise and identify the ITP-releasing neurons in the brain of L.maderae.

A well established method for the specific localisation of different substances in tissue sections is immunohistochemistry. It refers to the process of detecting antigens (e.g., proteins) in cells of a tissue section by exploiting the principle of antibodies binding specifically to antigens in biological tissues (Wikipedia article: Immunohistochemistry). Several procedures are used to obtain this type of information, based on high-affinity interactions between macromolecules, in the end leading to insoluble colored (or electron-dense) compounds, and so the localisation of specific substances with the help of light, and/or electron microscopy is possible (Junqueria et Carneiro, 2003).

For the localization of the ITP releasing neurons in the brain and periphery of the cockroach L.maderae, we applied two different indirect immunohistochemical methods. The first is the immunofluorescence staining and the second the peroxidase-antiperoxidase (PAP) staining method. As a primary antibody for the stainings we used rabbit antiCHH antibodies, which are polyclonal. Given that CHH has a very similar structure to ITP, an anti-crayfish CHH antibody can likely be used to stain cross-reactive ITP in the neurons of the L.maderae brain. The specificity of the primary antibody is however not high enough to distinguish between ITP and ITPL. Thus, in these experiments both splice forms are stained.

For the identification of ITP and ITPL in the nervous system of L.maderae we applied a high-performance liquid chromatography (HPLC) separation of extracts, followed by a direct enzyme-linked immunosorbent assay (ELISA) of the retrocerebral complex (corpora cardiaca and corpus allatum). We use the retrocerebral complex for the HPLC & ELISA, because many neurohaemal organ-like structures have been found within insects previously via immunoreactivity to CHH antibodies (Lucas, 2006).

## 2. Materials and Methods

### 2.1. Leucophaea maderae

During all experiments the cockroach Leucophaea maderae (L.maderae) was used as a model organism.

Adult cockroaches were taken from laboratory colonies kept under crowded conditions (at least 50 animals per 0,175m²) at room temperature in a fume hood. They were reared under a 12:12 h light-dark photoperiod. The animals were fed every 2-3 days with omnivore food, like meat, vegetables (e.g. eggplants, salad), fruits (e.g. apples, oranges), cheese and bread.

### 2.2. Dissection of Leucophaea maderae

### 2.2.1. Dissection

The animals were caught (one by one) using large forceps and put directly into a killing jar containing ethyl acetate vapours. The vapours killed the insects quickly without destroying them. After the cockroaches had died, the head was cut of with dissection scissors. The head was placed into a dissection bowl filled with Sylgard® plastics and Leucophaea-maderae ringer-solution (recipe in appendix). The head was dissected under a stereo-microscope and with the help of a pair of forceps and dissection scissors. The whole brain with the optic lobes, corpora cardiaca, corpora allata, sub-oesophageal ganglion, frontal ganglion and some peripheral nerves was dissected. The brain was largely freed from other tissues such as trachea, cuticule, muscles, mandibles, fat body etc.

### 2.2.2. Fixation

The fully dissected brains were fixed immediately by submersion. Therefore, the brains were placed on a piece of filter-paper, quickly blotted to near dryness and then put upside down into a lymph vessel filled with Zamboni's fixative (recipe in appendix). Because of being reasonably attached to the filter paper, the brains kept their original shape and connections. The active components in Zamboni's fixative are picric acid, which precipitates and derivatises proteins, and formaldehyde, which cross-links the proteins in the tissue, thus preserving the tissues from decay (Wikipedia article: Fixation (histology)). The brains were incubated in the fixative overnight at room temperature.

## 2.3. Cryostat Microtome Sectioning

### 2.3.1. Coating the Slides with Poly-L-Lysine

The Poly-L-Lysine solution (0.1% w/v, in water, containing Thimerosal, 0,01%, added as preservative, Sigma-Aldrich, Germany) was diluted to a final concentration to 0,01% with deionized water. The clean slides were placed in the diluted solution for 5 minutes. Afterwards they were drained and left to dry at room temperature overnight.

By the use of Poly-L-Lysine for the coating of the slides, the sections adhered more strongly to the glass surface.

### 2.3.2. Preparing and Embedding of the Brains

After the fixation of the dissected brains, the fixative and the filter-paper were removed, and the preparations were rinsed 3 times for 10 minutes in 0.1M phosphate buffer (0.1M PB) at room temperature. Subsequently, the preparations were incubated in a solution series with increasing sucrose concentrations, to introduce a freezing protectant before cryostat microtome sectioning. They were placed in each of the following sucrose-solutions for one and a half hour at room temperature under gently shaking: 5% sucrose, 10% sucrose and 20% sucrose in 0.1M PB. Each solution contained 0.02% sodium-azide.

Thereafter the preparations were embedded in freezing medium (Tissue Tek O.C.T. TM Compound): A drop of the freezing medium was placed on the metal tissue holder of the cryostat microtome and frozen in the cryostat microtome at -27°C. In a next step, again a small amount of freezing medium was put on the covered metal tissue holder and the sucrose-infiltrated brain was placed into it. The sample was then immediately inserted into the cryostat microtome at -27°C to allow for further freezing and equilibration.

### 2.3.3. Sectioning

For the sectioning, the cryostat microtome Leica CM 1850 UV, Germany, was used. The freezing chamber was cooled down on -27°C and the sections had a thickness of 20µm. Consecutive sections were collected on Poly-L-Lysine slides, with the help of a small paint brush. After sectioning, the slides were left in the freezing chamber of the cryostat microtome at -27°C overnight.

## 2.4. Vibratome Sectioning

### 2.4.1 Embedding

After the brains were fixed in Zamboni's fixative overnight, the fixative and the filter-paper were removed. Subsequent, the brains were washed 4 times for 10 minutes with 0.1M PB under gentle shaking and

incubated for 10 seconds at room temperature in a 5% gelatin/0.1M PB solution. The gelatin/albumin embedding medium (recipe in appendix) was melted by heating slightly in a water bath to about 45-50°C. A teflon form ($1x1x1.5cm^3$) was filled with the medium, the brain was placed into it and orientated immediately. The embedding medium was then left to harden by keeping it at 4°C for 30 minutes. Subsequently, the form was placed into a 3.7%-formalin solution for 1h at 4°C to prefix the embedding medium. Then the blocks with the embedded brains were taken out of the form and fixed further in the 3.7%-formalin solution overnight at 4°C in a tightly stoppered container. Blocks were briefly washed and kept in 0.1M PB at 4°C.

### 2.4.2. Sectioning

The sectioning was done with a vibrating blade microtome (Vibratome Series 1000 Sectioning System, Bachofer, Germany). While being sectioned, the preparations were immersed in a 0.1M phosphate buffered saline (0.1M PBS): The embedded preparations were cut with a razor blade in a trapezoidal shape and then glued with instant adhesive to a metal block. The block was placed and clamped into the holder of the vibratome. The preparations were sectioned at a speed of 2 and an amplitude of 4.5; every section had a thickness of 50µm. To store and later wash the sections, they were kept free-floating in 0.1M PB in sterile 24-well-plates.

## 2.5. Immunohistochemistry: Immunofluorescence Staining

Whole mount preparations or frozen sectiones were labelled using immunofluorescence staining. The same protocol was used for both, but the whole mount preparations were stained and washed in watch glasses and the frozen sections directly on the Poly-L-Lysine-coated slides.

### 2.5.1. Antibody-labelling

To prepare whole mounts and section of brains for the labelling with the primary antibody, samples were washed 3 times for 10 minutes with 0.1M PB and then 3 times for 10 minutes with Tris-buffered saline, containing 0.1% Triton X-100 (0.1M TBTX), to remove most of the lipids from the tissue. Subsequently, they were incubated overnight at room temperature in the primary antibody solution. For the whole mount preparations, the primary antibody antiOrconectes CHH T5 B4/70 (1:5000) in 0.1M TBTX, containing 0.02% sodium-azide, was used. For the frozen sections, the primary antibody anti-OCHH T5 B/4 (1:8000) in 0.1M TBTX, containing 0.02% sodium-azide, was used. The primary antibody targets the antigen ITP in the tissue.

After the incubation, the primary antibody solution was removed, and the brains were washed 3 times for 15 minutes with 0.1M TBTX and afterwards incubated for 2 hours at room temperature in the secondary antibody solution. For the whole mount preparations, the secondary antibody GAR-FITC (Goat anti Rabbit IgG (whole molecule) Fluorescein Isothiocyanate (FITC) Conjugate, Sigma-Aldrich, Germany) 1:100 in 0.1M TBTX, containing 0.02% sodium-azide, was used and for the frozen sections the secondary antibody GAR-Cy3 (Goat anti Rabbit, Cyanin 3) 1:1000 in 0.1M TBTX, containing 0.02% sodium-azide. The secondary antibody, which is loaded with a fluorescence marker (FITC or Cy3), binds specifically to the primary rabbit antibody. Thereafter the preparations were washed 3 times for 10 minutes with 0.1M TBTX.

The detailed protocol for the Antibody-labelling/Immunofluorescence staining can be found in the appendix (6.3.1).

### 2.5.2. Mounting

To mount the whole mount preparations, they were placed with 0.1M PB into the concave well in the middle of an hour glass slide; The buffer was then sucked off. A small amount (500µl) of a solution containing 80% glycerol (Merck) and 50mg/ml 1,4-Diazabicyclo[2.2.2]Octane (DABCO, Triethylenediamine, Sigma-Aldrich, Germany) was then pipetted onto the whole mounts or frozen section on the slides. Cover slips (size: 24 x 50mm$^2$) were finally placed on the slides.

The preparations were then ready for being viewed under the fluorescence microscope (Zeiss Axioplan 2 or Leitz Aristoplan).

### 2.6. Immunohistochemistry: Peroxidase-Antiperoxidase Staining

### 2.6.1. Antibody-labelling and Peroxidase-Antiperoxidase-Complex

After the Vibratome sectioning, the free-floating sections were washed 3 times for 10 minutes with 0.1M TBTX and then preincubated in 2% NGS in 0.1M TBTX, containing 0.02% sodium-azide, for 1 hour at room temperature. Subsequently, they were incubated in a primary antibody solution at room temperature overnight continous gently shaking. The primary antibody anti-OCHH T5 B/4 (1:8000) in 0.5M TBTX, containing 1% NGS and 0.02% sodium azide, was used.

Afterwards, the sections were rinsed 3 times for 10 minutes in 0.1M TBTX. The second antibody labelling was done with GAR (Goat anti Rabbit IgG without any marker, ICN ImmunoBiologicals) (1:100) in 0.1M TBTX, containing 1% NGS and 0.02% sodium-azide, for 1 hour at room temperature under gentle shaking. The sections were washed 3 times for 10 minutes with 0.1M TBTX after the second antibody labelling. Thereafter, the sections were incubated in rabbit PAP (1:300) in 0.5M TBTX, containing 1% NGS, for 1 hour at room temperature while gently shaking. The rabbit-PAP-comples, consisting of three peroxidase and two anti-peroxidase molecules, binds to the secondary antibody GAR (Sternberger, 1974). Subsequently, they were washed 3 times for 10 minutes with 0.1M TBTX and additionally one time for 10 minutes with 0.1M PB.

### 2.6.2. Diaminobenzidine-reaction

Diaminobenzidine(DAB) solutions were always freshly made before use. 10 mg DAB powder (3,3'-Diaminobenzidine tetrahydrochloride hydrate, >96%, Sigma-Aldrich, Germany) was first diluted in 500µl distilled water and then added to 30ml 0.1M PB, containing 10µl $H_2O_2$ (from 30% stock). The DAB solution was poured onto the PAP-labelled sections. DAB is oxidized in the presence of peroxidase and a brown precipitate is produced consequently, which can be seen under a light (or electron) microscope (Junqueira et Carnerio, 2003). The reaction time was 10-45 minutes under visual control with a stereo-microscope. As soon as the sections showed a brown background staining, the reaction was stopped. Thereafter, the sections were rinsed 3 times for 2 minutes in distilled water. Finally, they were washed once in 0.1M PB and also kept in 0.1M PB at 4°C.

The detailed protocol for the PAP method is given in the appendix.

### 2.7. Permanent Preparations

### 2.7.1. Coating the Slides with chromalum-gelatin

The slides were cleaned in a acetone/ethanol(1:1) solution overnight. The chromalum-gelatin solution (recipe see appendix) was always freshly prepared. The slides were put one by one put for 10 seconds into the chromalum-gelatin solution, slowly removed and afterwards leaned against a vessel to drain. The slides were then placed in a vessel and dried overnight at 60°C.

### 2.7.2. Placing sections on coated slides

To place sections on a chromalum-gelatin coated slide, a small amount of 0.1M PB was pipetted on the slide. Consecutive sections were then positioned one by one on the slide and left to dry at room temperature. A small amount of a 2% gelatin-solution was finally placed on top of the sections in order to avoid folding up during drying, which later proceeded at room temperature overnight.

### 2.7.3. Dehydration and Mounting

For permanent preparations, the sections were dehydrated in an increasing alcohol series and cleared in xylene. All solutions in this procedure were provided in 250ml cuvettes, and all incubations were done at room temperature. The slides with the sections were incubated for 10 minutes in 0.1M PB and then for 3 minutes in distilled water. Accordingly, the slides were incubated for 3 minutes in each of the following alcohol concentrations: 50%, 60%, 70%, 75%, 80%, 85%, 90%, 95%, 100%, 100%. Afterwards, the slides were cleared 3 times for 3 minutes in xylene (3 different vessels). Subsequently, a small amount (1-2ml) of DEPEX (DPX mountant for microscopy (contains xylene (mixture of isomers), dibutyl phthalate, BDH laboratory supplies, England) was placed on each slide and a cover slip on top. They were then left to dry at room temperature overnight.

The permanent preparations were then viewed using bright-field optics under a Leitz Aristoplan microscope.

### 2.8. High-Performance Liquid Chromatography (HPLC)

### 2.8.1. Dissection for HPLC

Animals were sacrificed as described under 3.2.1 Dissection. For high-performance liquid chromatography (HPLC) separation only the retrocerebral complexes (RCC) (corpora cardiaca and corpora allata) were dissected. These dissections were quickly done under a stereo-microscope in a dissection bowl filled with Leucophaea-maderae ringer-solution. Finished preparations were collected on the bottom of dry ice-chilled 1,5 ml Eppendorf tubes and finally kept at -80°C.

### 2.8.2. Preparation for HPLC: Acidic Extraction

At first, 500µl of 2M acetic acid were pipetted onto 10 L.maderae RCCs in the Eppendorf tube (1,5ml). This solution was treated with a ultrasonic disruptor 3 times for 30 seconds on ice. Afterwards, it was centrifuged for 10 minutes at 16.000g at 4°C. The supernatant was pipetted out and saved in a new Eppendorf tube (1,5ml). Again 500µl of 2M acetic acid was poured on the pellett, and the new solution was treated with the ultrasonic disruptor 3 times for 30 seconds on ice. Subsequently, the solution was centrifuged a second time for 10 minutes at 16.000g at 4°C. The supernatant was transferred together with the supernatant from the first centrifugation. Then, the solution was dried in a vacuum centrifuge (SpeedVac SC100, Savant coupled to a Freeze Dryer 4.5 Labconco; Pfeiffer Vacuum Pump) for ca. 1 hour at -50°C. Afterwards, the fractions were resuspended with 200µl 2M acetic acid and again centrifuged for 10 minutes at 16.000g at 4°C before injection into the HPLC.

### 2.8.3. HPLC

The reversed-phase HPLC fractionation of the acidic extracts was performed on a Waters HPLC system with a U6K injector, a model 600E controller, and a model 486 ultraviolet detector (set to 210nm). A Waters Millennium 2010 chromatography manager was used for data acquistion and processing. After 2 minutes of initial conditions, a linear gradient was applied from 18% to 48% acetonitrile containing 0.1% Trifluoroacetic acid (TFA) for 60 minutes (MeCN-TFA; 0,5% MeCN/minute) followed by a final 10-minute gradient to 60% MeCN-TFA at a flow rate of 0.9 ml/minute. During chromatography on a µ-Bondapak phenyl column (4,6 mm x 250 mm; Waters; with a guard column) peak fractions were collected manually (Dircksen et al. 2008).

Afterwards, the fractions were dried in the vacuum centrifuge for 1 hour at -50°C and then kept at -20°C.

### 2.9. Enzyme-Linked Immunosorbent Assay (ELISA)

### 2.9.1. Antigen loading

At first, 100µl of 60% acetonitrile was pipetted on every dry fraction, that was then vortexed for 10 seconds, put in an ultrasonic bath (Sonorex RK 255 S, Bandelin) for 15 seconds and then vortexed again for 10 seconds.

From every HPLC fraction and a blank sample, 10µl were loaded in 2 wells each of a sterile 96-well-plate (pipetting scheme in appendix). The plate was then dried under vacuum for half an hour at -50°C.

Subsequently, 100µl of 0.1M sodium-carbonate buffer (pH 9.6) were added to every well, and the plate was then incubated for 2 hours at 37°C in a humid chamber.

### 2.9.2. Primary antiserum loading

The sodium-phosphate buffer was chucked out from the 96-well-plate and all wells washed 3 times for 5 minutes with 150µl PBS-Tween-azide (see appendix for buffer details) per well, while shaking at 450rpm. Subsequently, 100µl of the primary antiserum solution (antiOCHH T5 B/4 (1:8000) in PBS-Tween-azide, containing 1% NGS) was pipetted in each well. The samples were incubated overnight at 4°C in the primary antiserum.

### 2.9.3. Goat Anti Rabbit Alkaline Phosphatase Conjugate

The primary antiserum was chucked out from the 96-well-plate, and the samples in the wells were again washed 3 times for 5 minutes with 150µl PBS-Tween-azide per well under rapid shaking (shaker agitation 450 rpm).

Then 100µl of 1% goat anti rabbit alkaline phosphatase (GAR-AP) in PBS-Tween-azide, containing 1% NGS, was pipetted in each well, and incubated for 1,5 hours at 37°C in a humid chamber.

### 2.9.4. Substrate solution loading and photometric measurements

The GAR-AP solution was chucked out from the plate, and the samples were washed again 3 times for 5 minutes with 150µl PBS-Tween (without azide), under agitation of 450rpm. Afterwards, 100µl of  the substrate solution consisting of 0.1% para-nitrophenyl phosphate (4-Nitrophenylphosphate Disodium salt, Hexahydrate, >97% (NT), Fluka Biochemika, Germany) in Na-carbonate buffer pH 9.6) were pipetted into each well. After 15, 30 and 45 minutes, the absorption of the samples were measured with an ELISA reader (Titertek Multiskan Plus MK 2).

### 3. Results

### 3.1. Immunofluorescence Staining

### 3.1.1. Immunofluorescence stained whole mount preparations

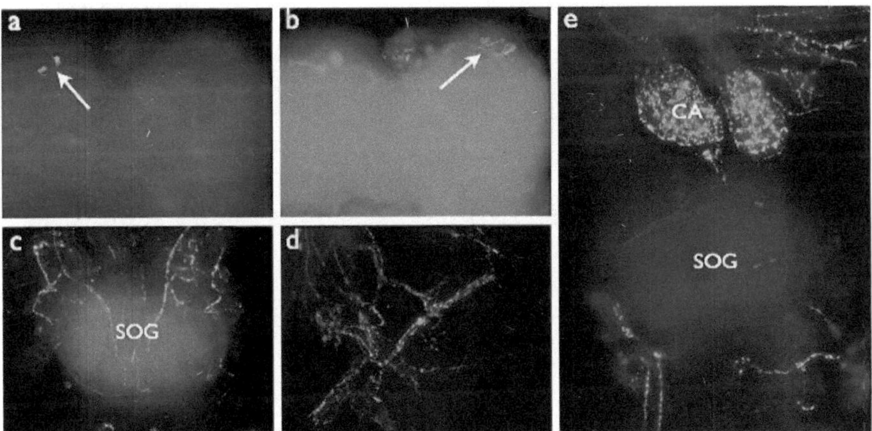


Fig. 1. Immunohistochemical localisation of the Ion transport peptide in whole mount preparations of the brain of Leucophaea maderae. (a-e) The preparations are stained with the primary antibody antiCHH and the secondary antibody GAR-FITC. (a-b) In the protocerebrum of the brain of L.maderae two cell groups with a immunoreactivity to the antiCHH antibody are shown. They are localisad in the frontal part, one group on each

side (arrows). (c-e) The sub-oesophageal ganglion (SOG) of the L.maderae brain is shown. Many structures, which are immunoreactive to the antiCHH antibody, occur in the periphery around the SOG nerves and nerve roots. (e) Fibre networks in the corpus allatum (CA), heavily stained by the antiCHH antibody, can be seen above the SOG. Two bilaterally symmetrical cell groups occur in the frontal part of the protocerebrum of the brain, which showed an immunoreactivity to the antiOCHH antibody (Fig. 1a and 1b, marked by the arrows).

Furthermore, we discovered a lot of fibres in the periphery around the sub-oesophageal ganglion, which were heavily stained by the antiCHH antibody (fig. 1c and 1d). The retrocerebral complex also showed a strong immunreactivity to the antiCHH antibody, especially the corpus allatum showed abundant fibres and terminals of an intense fluorescence ocurred around the glandular cells (Fig. 1e). Comparatively less abundant fibres and terminals were found in the corpora cardiaca.

## 3.1.2. Immunofluorescence stained cryostat sections

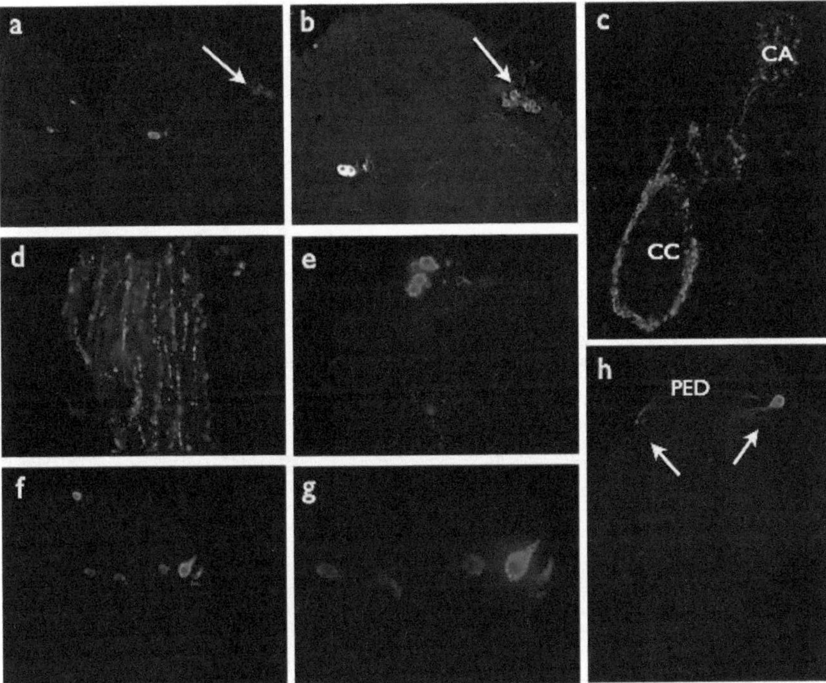

Fig. 2. Immunohistochemical localisation of Ion transport peptide in longitudinal cryostat sections of the brain of Leucophaea maderae. (a-h) The sections are stained with the primary antibody antiOCHH T5 B/4 and the secondary antibody GAR-Cy3. (a-b) Sections of protocerebrum. A group of 6 cells occurs in an anterior lateral position (arrows). (c) CHH-immunostained retrocerebral complex (copus cardiacum CC, corpus allatum CA). (d) Muscle cells in the periphery of the brain, showing varicose fibres running along the muscle fibres. (e) A group of 4-5 cells in the lateral protocerebrum. (f-g) A group of 6 cells at higher magnifications. (h) A neuron wihtin the superior lateral protocerebrum with its dendrites (arrows), projecting to the median bundle (MB) (pedunculs PED).

We found a second immunoreactive cell group, consisting of 6 cells, localised in the frontal lateral part of the protocerebrum, lying closer to the optic lobes than the first cell group, described in 4.1.1 Immunofluorescence stained whole mount preparations (Fig. 2a and 2b, marked by the arrows).

Moreover, we saw many structures showing immunoreactivity to the antiOCHH antibody around unidentified cross-striated muscle cells in the periphery around the brain (fig. 2d). We were also able to find stained neurons near the pedunculus of mushroom bodies in the L.maderae brain. The dendrites of the neurons extend around the pedunculus, but are not connected with it (Fig. 2h).

As already seen in the whole mount preparations, fibres in the corpus allatum showed a strong staining (Fig. 2c).

## 3.2. Peroxidase-antiperoxidase Staining

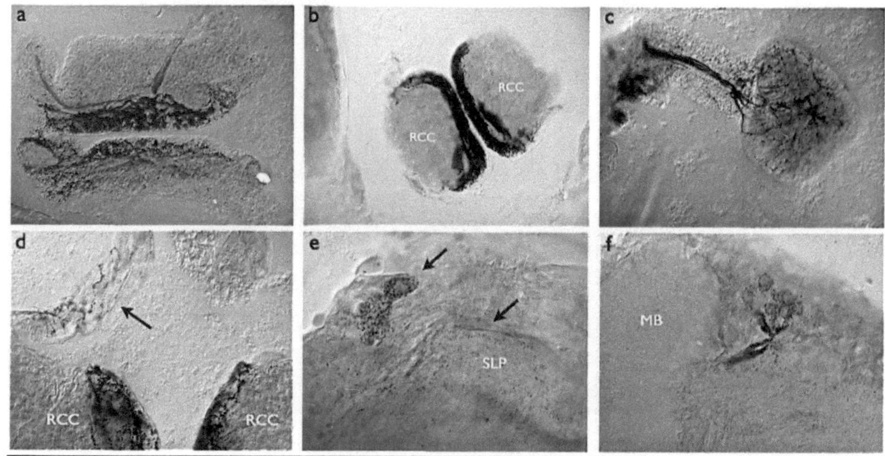

Fig. 3. Immunohistochemical localisation of Ion Transport Peptide in longitudinal vibratome sections of the brain of the cockroach Leucophaea maderae. (a-h) The sections are stained with the primary antibody antiOCHH T5 B/4, the secondary antibody GAR and a PAP-complex. (a) CHH-immunostained RCC (longitudinal). (b) CHH-immunostained RCC (transversal). (c) Section of the CA, showing a nerve bundle inside. (d) Immunoreactive, cross-striated muscle cells near the RCC (arrow). (e) A group of cells near the SLP. (f) A group of cells with their axons, projecting to the MB. Examinating the PAP-stained vibratome sections, we found several CHH-immunoreactive structures. As already shown by the immunofluorescence-stained whole mount preparations and cryostat sections, the RCC appeared to be immunoreactive to the antiOCHH antibody. Especially in the middle part of the RCC, where the corpara cardiaca meet, a strong staining could be recognized (Fig. 3a-b). We were able to show a bundle of nerves along the CA and cellbodies at the end of the CA, which are CHH-immunoreactive (Fig. 3c).

In the periphery near to the RCC we found a strand of unidentified cross-striated muscle cells, surrounded by many structures, showing a immunoreactivity to the antiOCHH antibody (Fig. 3d, arrow).

Furthermore we found CHH-immunoreactive cell groups with their axons, projecting to the SLP/MB, in the protocerebrum (Fig. 3e-f, arrows).

### 3.3. HPLC and ELISA

### 3.3.1. HPLC

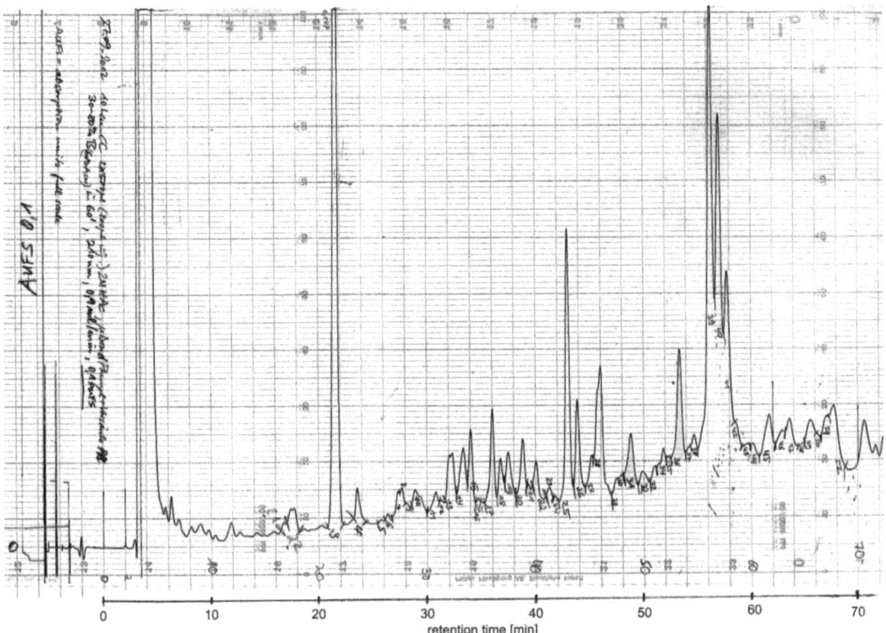

Fig. 4. Graphical representation of a reversed-phase HPLC of a 0.2M acetic acid extract of 10 dissected RCCs from L.maderae on a Waters µBondapak phenyl column. The yellow peaks represent the substances, which had the highest absorbance in the ELISA testing, which was done later on.

By the use of a high-performance liquid chromatography of 10 retrocerebral complexes of the cockroach L.maderae, we were able to separate 52 substances (shown by the peaks in the graphic Fig.4). The higher the retention time is, the more hydrophobic are the substances.

The two yellow marked peaks represent the substances, which later on had the highest absorbance in the ELISA testing.

### 3.3.2. ELISA

**Direct ELISA** 405nm  from 03.10.12;  10 LemRCCs from 27.09.2012

2NHac extract HPLC uBondPh  30-80%B60, antiOCHH 1:8k

2. Measurement: after 30 minutes at 17:14:44     Arithmetic average and blank adjusted

|   | 1 | 2 | 3 | 4 | 5 | 6 | 7 | 8 | 9 | 10 | 11 | 12 |
|---|---|---|---|---|---|---|---|---|---|----|----|----|
| A | 0,000 | 0,000 | 0,000 | 0,000 | 0,046 | 0,046 | 0,000 | 0,000 | 0,021 | 0,021 | 0,000 | 0,000 |
| B | 0,000 | 0,000 | 0,050 | 0,050 | 0,000 | 0,000 | 0,004 | 0,004 | 0,000 | 0,000 | 0,037 | 0,037 |
| C | 0,032 | 0,032 | 0,080 | 0,080 | 0,000 | 0,000 | 0,002 | 0,002 | 1,581 | 1,581 | 0,000 | 0,000 |
| D | 0,000 | 0,000 | 0,000 | 0,000 | 0,000 | 0,000 | 0,000 | 0,000 | 0,040 | 0,040 | 0,000 | 0,000 |
| E | 0,006 | 0,006 | 0,000 | 0,000 | 0,000 | 0,000 | 0,006 | 0,006 | 0,003 | 0,003 | 0,003 | 0,003 |
| F | 0,000 | 0,000 | 0,000 | 0,000 | 0,048 | 0,048 | 0,536 | 0,536 | 0,000 | 0,000 | 0,000 | 0,000 |
| G | 0,000 | 0,000 | 0,000 | 0,000 | 0,000 | 0,000 | 0,000 | 0,000 | 0,000 | 0,000 | 0,000 | 0,000 |
| H | 0,000 | 0,000 | 0,003 | 0,003 | 0,000 | 0,000 | 0,000 | 0,000 | 0,000 | 0,000 | 0,000 | 0,000 |

Fig. 5. Direct ELISA data of manually collected peaks from 10 dissected RCCs from L.maderae, after 30 minutes of incubation. Staining was done with a antiOCHH antibody (1:8000). Yellow marked boxes show the substances with the highest light absorbance.

For the direct ELISA procedure we used 47 (no. 3 - 49) of the before manually collected peaks from the HPLC of 10 RCCs of L.maderae. After the incubation, most of the substances showed a very low light absorbance (nearly zero), but two samples possessed to have a higher light absorbance. These were the samples no. 30 and 36 (yellow marked in Fig.5), which are also yellow marked in the graphical representation of the HPLC (Fig.4). They had a absorbance of 0,536 (no. 30) and 1,581 (no. 36).

### 4. Conclusions and Perspectives

During my work I made a lot of experiences with the different techniques and can now give some suggestions how the best results can be achieved and compare the different techniques with each other.

To localise the ITP producing and/or releasing cells in the brain of the cockroach L.maderae by using the fluorescence staining, it is better to utilise sections, cutted with microtomes, than using whole mount preparations. The whole mount preparations are much too thick, to gain strong stained cells, which are located deep in the protocerebrum.

For the PAP staining and the following generating of permanent preparations it is best to use vibratome sections. By using the frozen sections the risk of losing the sections, while performing the many washing steps or incubating with antibody solutions or alcohol, is too high.

The benefit ot the PAP staining is primarily the high binding significance and the strong staining, even though the first antibody is used in very low concentrations. But the PAP staining takes much more time than the fluorescence staining.


The results of the HPLC and the following ELISA are showing us, that there are 2 substances in the RCC of the cockroach L.maderae, which are immunoreactive to the antiCHH antibody. Based on the knowledge we have now, the neurohormone CHH (not CHH-family peptides) only occurs in crustaceans (Webster et al, 2011). In some insects (like locusts) exists the neurohormone ITP, which has a lot of similarities in structure and function with CHH (Dircksen, 2008). Therefor we can suggest that the immunoreactive substances are ITP or at least that the samples contain ITP (and not CHH). So we can say, that we identified ITP in the RCC of L.maderae.

By the use of the immunohistochemical techniques (immunofluorescence staining and PAP staining), we were able to localise a lot of structures and cells directly in the brain tissue and the surrounding periphery of the cockroach L.maderae, which are immunoreactive to the antiCHH and antiOCHH antibody. In total we found 2 parallel cell groups in the protocerebrum, a lot of structures in the RCC and around the muscle cells, that are producing and/or releasing the neurohormone ITP.

So the conclusion is, that L.maderae maybe controlls processes like osmoregulation, moulting and reproduction with the help of ITP.

In the SLP of the cockroach brain we found CHH immunoreactive neurons. The dendrites of the neurons extend around the pedunculus, but are not connected with it. The pedunculus is part of the mushroom body, which is responsible for higher, integrative performances like learning and memorising (Wikipedia article: Mushroom bodies). The pedunculus and the CHH immunoreactive neuron do not feature a connection between each other. This indicates, that no memory is needed for the production of ITP.

CHH is a hyperglycaemic hormone in crustaceans, which has a central role in blood glucose regulation (Lorenzon, 2005). As metioned above, ITP has a lot of similarities with CHH, and so it is possible, that ITP is also one of the regulating factors in terms of blood glucose in insects. We found a lot of CHH antibody immunoreactive structures surrounding the muscle cells of L.maderae. As insects do not have a liver like mammals, it can be, that the muscle cells of the insects realease sugar into the haemolymph with the help of ITP to regulate the „blood" glucose.

We were able to show that a CHH like substance occurs in the cockroach L.maderae and where it is located, but nevertheless further studies are needed to prove that this substance is really ITP. Also some reproductive work is required, because we found two CHH immunoreactive samples from the RCC by the use of HPLC and ELISA. These substances can be the two isoforms of ITP or a mistake was done during the dissection of the RCCs or the manual collection of the HPLC samples.
Moreover the function of ITP in the cockroach L.maderae is completly unexplored, yet. To discover those, more studies will be necessary.

# 5. List of Abbrevations

| | |
|---|---|
| CA | Corpus allatum |
| CC | Corpus cardiaca |
| CHH | Crustacean Hyperglycaemic Hormone |
| DAB | 3,3-Diaminobenzidine |
| DABCO | DABCO |
| ELISA | Enzyme Linked Immunosorbent Essay |
| GAR | Goat Anti Rabbit |
| - FITC | - Fluorescein Isothiocyanate |
| - Cy3 | - Cyanin 3 |
| GAR-AP | Goat Anti Rabbit Alkaline Phophatase |
| HPLC | High Performance Liquid Chromatography |
| IHC | Immunohistochemistry |
| ITP | Ion Transport Peptide |
| ITPL | Ion Transport Peptide Long Splice Form |
| L.maderae | Leucophaea maderae |
| MB | Median Bundle |
| NGS | Normal Goat Serum |
| PAP | Peroxidase-Antiperoxidase |
| PB | Phosphate Buffer |
| PBS | Phosphate Buffered Saline |
| PED | Pedunculus |
| RCC | Retrocerebral Complex |

| SLP | Superior Lateral Protocerebrum |
| SOG | Sub-Oesophageal Ganglion |
| TBTX | Tris Buffered Triton X-100 |

## 6. Recipes

<u>0.1M Phosphate Buffer (PB) (0.1M Na-phosphate buffer, pH 7,4)</u>

1:1 diluted from 1 litre of stock solution (0,2M PB):


28,045 g Na2HPO4 x 2 H2O

5,244 g   NaH2PO4 x 1 H2O


dissolved in 1 litre of bidistilled H2O


<u>0.1M TBTX (0,3M NaCl, 0.1M Tris/HCl, 0.1% Triton X-100, pH7,4)</u>

17,532 g NaCl

12,114 g TrisBase

1 ml Triton X-100


dissolved in 1 litre of bidistilled H2O


<u>0.5M TBTX (0,3M NaCl, 0.1M Tris/HCl, 0,5% Triton X-100, pH 7,4)</u>

1 litre TBTX 0,1

4 ml Triton X-100


<u>Chromalum-gelatin</u>

1 g gelatin

0,1 g kaliumchrom(III)sulfat


dissolved in 100 ml bidistilled H2O


<u>Embedding medium (gelatin/albumine)</u>

1. Dissolve 75g Ovalbumin (Sigma A 5253 grade II) in 200ml distilled water (at least 1 hour)
2. Heat 50ml distilled water up to 50°C and dissolve 11g gelatin (Sigma G 2500 300 bloom) in it.
3. Add ca. 80 Vol % Ovalbumin solution to the warm gelatin solution.
4. Add 0,05% $NaN_3$.

Leucophaea-maderae ringer-solution (pH 6,8 - 7,0)

9 g NaCl

0,2 g KCl

0,265 g CaCl2 x 2 H2O

4 g Glucose

3,12 g Hepes


dissolved in 1 litre of bidistilled H2O


Na-carbonate Buffer (pH 9.6)

| 0,62g / 50ml   $Na_2CO_3$ | 29,3ml |
| 0,84g / 100ml NaHCO$_3$ | 70,7ml |
|  | 100 ml |


PBS-Tween-azide

1l  0.1M PB (pH 7.4),

plus 0,9 % NaCl,

plus 0.1% Tween20,

plus 0.02% Na-azide


Phosphate buffered saline (PBS)

50 ml 0,2M PB

950 ml bidistilled H2O

9 g NaCl


Zamboni's Fixative

1)dissolve 4g paraformaldehyde in 50ml H20 at 65°C, clear with 1-2 drops 10N NaOH

2)dissolve 0,4g NaH2PO4 x 1 H2O and 0,815g NaHPO4 x 2 H2O in 40ml distilled
   water

agitate these two solutions with 7,5ml saturated aqueous picric acid and fill ad 100ml

## 7. Principles

### 7.1. Immunohistochemistry:

Antibody-types:

The antibodies used for specific detection can be polyclonal or monoclonal. Polyclonal antibodies are made by injecting animals with peptide antigen and, after a secondary immune response is stimulated, isolating antibodies from whole serum. Thus, polyclonal antibodies are a heterogeneous mix of antibodies that recognize several epitopes. Monoclonal antibodies show specificity for a single epitope and are therefore considered more specific to the target antigen than polyclonal antibodies.

For IHC detection strategies, antibodies are classified as primary or secondary reagents. Primary antibodies are raised against an antigen of interest and are typically unconjugated (unlabelled), while secondary antibodies are raised against immunoglobulins of the primary antibody species. The secondary antibody is usually conjugated to a linker molecule, such as biotin, that then recruits reporter molecules, or the secondary antibody itself is directly bound to the reporter molecule (Wikipedia article: Immunohistochemistry).

Immunohistochemistry (IHC) reporters:

Reporter molecules vary based on the nature of the detection method, and the most popular methods of detection are with enzyme- and fluorophore-mediated chromogenic and fluorescence detection, respectively. With chromogenic reporters, an enzyme label is reacted with a substrate to yield an intensely colored product that can be analyzed with an ordinary light microscope. While the list of enzyme substrates is extensive, Alkaline phosphatase (AP) and horseradish peroxidase (HRP) are the two enzymes used most extensively as labels for protein detection. An array of chromogenic, fluorogenic and chemiluminescent substrates is available for use with either enzyme, including DAB or BCIP/NBT, which produce a brown or purple staining, respectively, wherever the enzymes are bound.

Reaction with DAB can be enhanced using nickel, producing a deep purple/black staining. Fluorescent reporters are small, organic molecules used for IHC detection and traditionally include FITC, TRITC and AMCA. For chromogenic and fluorescent detection methods, densitometric analysis of the signal can provide semi- and fully quantitative data, respectively, to correlate the level of reporter signal to the level of protein expression or localization (Wikipedia article: Immunohistochemistry).

Target antigen detection methods:

The indirect method involves an unlabeled primary antibody (first layer) that binds to the target antigen in the tissue and a labeled secondary antibody (second layer) that reacts with the primary antibody. The secondary antibody must be raised against the IgG of the animal species in which the primary antibody has been raised. This method is more sensitive than direct detection strategies because of signal amplification due to the binding of several secondary antibodies to each primary antibody if the secondary antibody is conjugated to the fluorescent or enzyme reporter (Wikipedia article: Immunohistochemistry).

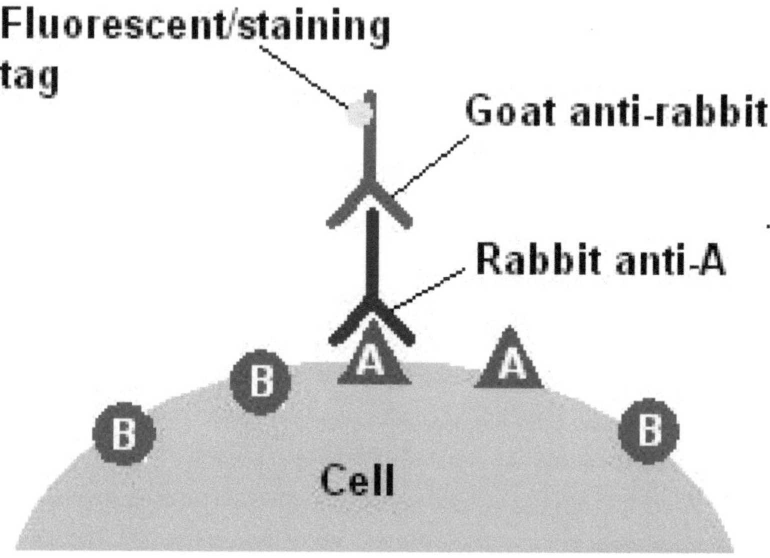


Fig. 6: The indirect method of immunohistochemical staining uses one antibody against the antigen being probed for, and a second, labelled, antibody against the first. (Wikipedia)

The indirect method, aside from its greater sensitivity, also has the advantage that only a relatively small number of standard conjugated (labeled) secondary antibodies needs to be generated. For example, a labeled secondary antibody raised against rabbit IgG, which can be purchased "off the shelf," is useful with any primary antibody raised in rabbit. With the direct method, it would be necessary to label each primary antibody for every antigen of interest (Wikipedia article: Immunohistochemistry).

## 7.2. Fluorescence

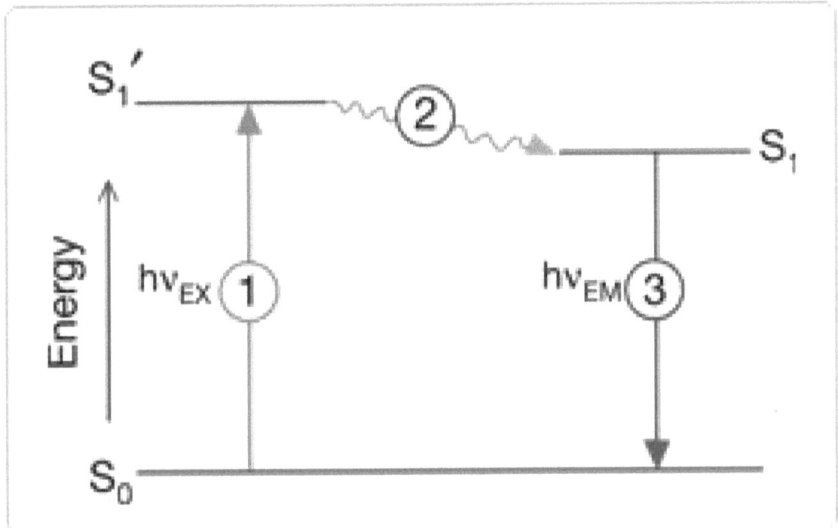


Fig. 7: Jablonski diagram illustrating the processes involved in the creation of an excited electronic singlet state by optical absorption and subsequent emission of fluorescence. The labeled stages 1, 2 and 3 are explained below (Haugland, 2002).

Stage 1: Excitation
A photon of energy $hv_{EX}$ is supplied by an external source such as an incandescent lamp or a laser and absorbed by the fluorophore, creating an excited electronic singlet state ($S_1'$). This process distinguishes fluorescence from chemiluminescence, in which the excited state is populated by a chemical reation (Haugland, 2002).

Stage 2: Excited-State Lifetime
The excited state exists for a finite time (typically 1-10 nanoseconds). During this time, the fluorophore undergoes conformational changes and is also subject to a multitude od possible interactions with its molecular environment. THese processes have two important consequences. First, the energy of $S_1'$ is partially dissipated, yielding a relaxed singlet excited state ($S_1$) from which fluorescence emission originates. Second,

not all the molecules initally excited by absorption (Stage 1) return to the ground state $(S_0)$ by fluorescence emission. Other processes such as collisional quenching, fluorescence resonance energy transfer and intersystem crossing (*see* below) may also depopulate $S_1$. The fluorescence quantum yield, which is the ratio of the number of fluorescence photons emitted (Stage 3) to the number of photons absorbed (Stage 1), is a measure to the relative extent to which these processes occur (Haugland, 2002).

Stage 3: Fluorescence Emission

A photon of energy $hv_{EM}$ is emitted, returning the fluorophore to tis ground state $S_0$. Due to energy dissipation during the excited-state lifetime, the energy of this photon i lower, and therefore of longer wavelength, than the excitaion photon $hv_{EX}$. The difference in energy or wavelength represented by $(hv_{EX}-hv_{EM})$ is called the Stokes shift. The Stokes shift is fundamental to the sensivitiy of fluorescence techniques because it allows emission photons to be detected against a low background, isolated from excitation photons. In contrast, absorption spectrophotometry requires measurement of transmitted light relative to high incident light levels at the same wavelength (Haugland, 2002).

**7.3. Reversed-phase High-performance liquid chromatography**

Reversed phase HPLC (RP-HPLC) has a non-polar stationary phase and an aqueous, moderately polar mobile phase. One common stationary phase is a silica which has been surface-modified with $RMe_2SiCl$, where R is a straight chain alkyl group such as $C_{18}H_{37}$ or $C_8H_{17}$. With such stationary phases, retention time is longer for molecules which are less polar, while polar molecules elute more readily.

RP-HPLC operates on the principle of hydrophobic interactions, which originates from the high symmetry in the dipolar water structure and plays the most important role in all processes in life science. RP-HPLC allows the measurement of these interactive forces. There is a proportionality between the analyte's stationary

phase and contact surface area around the non-polar segment of the analyte molecule after it's association with the ligand.

This solvophobic effect is dominated by the force of water for "cavity-reduction" around the analyte and the $C_{18}$-chain versus the complex of both. The energy released in this process is proportional to the surface tension of the eluent (water: $7.3 \times 10^{-6}$ J/cm², methanol: $2.2 \times 10^{-6}$ J/cm²) and to the hydrophobic surface of the analyte and the ligand respectively. The retention can be decreased by adding a less polar solvent (methanol, acetonitrile) into the mobile phase to reduce the surface tension of water. Gradient elution uses this effect by automatically reducing the polarity and the surface tension of the aqueous mobile phase during the course of the analysis.

Structural properties of the analyte molecule play an important role in its retention characteristics. In general, an analyte with a larger hydrophobic surface area (C-H, C-C, and generally non-polar atomic bonds, such as S-S and others) is retained longer because it is non-interacting with the water structure. On the other hand, analytes with higher polar surface area (conferred by the presence of polar groups, such as -OH, -$NH_2$, $COO^-$ or -$NH_3^+$ in their structure) are less retained as they are better integrated into water. Such interactions are subject to steric effects in that very large molecules may have only restricted access to the pores of the stationary phase, where the interactions with surface ligands (alkyl chains) take place. Such surface hindrance typically results in less retention.

Retention time increases with hydrophobic (non-polar) surface area. Branched chain compounds elute more rapidly than their corresponding linear isomers because the overall surface area is decreased. Similarly organic compounds with single C-C-bonds elute later than those with a C=C or C-C-triple bond, as the double or triple bond is shorter than a single C-C-bond.

Another important factor is the mobile phase pH since it can change the hydrophobic character of the analyte. For this reason most methods use a buffering agent, such as sodium phosphate, to control the pH. Buffers serve multiple purposes: control of pH, neutralize the charge on the silica surface of the stationary phase and act as ion pairing agents to neutralize analyte charge. Ammonium formate is commonly added in mass spectrometry to improve detection of certain analytes by the formation of analyte-ammonium adducts. A volatile organic acid such as acetic acid, or most commonly formic acid, is often added to the mobile phase if mass spectrometry is used to analyze the column effluent. Trifluoroacetic acid is used infrequently in mass spectrometry applications due to its persistence in the detector and solvent delivery system, but can be effective in improving retention of analytes such as carboxylic acids in applications utilizing other detectors, as it is a fairly strong organic acid. The effects of acids and buffers vary by application but generally improve chromatographic resolution (Wikipedia article: HPLC).


## 7.4. Enzyme-Linked Immunosorbent Assay

As a "wet lab" analytic biochemistry assay, ELISA involves detection of an "analyte" in a liquid sample by a method that continues to use liquid reagents during the "analysis" that stays liquid and remains inside a reaction chamber or well needed to keep the reactants contained.


As a heterogenous assay, ELISA separates some component of the analytical reaction mixture by adsorbing certain components onto a solid phase which is physically immobilized. In ELISA, a liquid sample is added onto a stationary solid phase with special binding properties and is followed by multiple liquid reagents that are sequentially added, incubated and washed followed by some optical change in the final liquid in the well from which the quantity of the analyte is measured. The qualitative "reading" usually based on detection of intensity of transmitted light by spectrophotometry, which involves quantitation of transmission of some specific wavelength of light through the liquid (as well as the transparent bottom of the well in the multiple-well plate format). The sensitivity of detection depends on amplification of the signal during the analytic reactions. Since enzyme reactions are very well known amplification processes, the signal is generated by enzymes which are linked to the detection reagents in fixed proportions to allow accurate quantification (Bovine OPG).

The ligand-specific binding reagent is "immobilized", i.e., usually coated and dried onto the transparent bottom and sometimes also side wall of a well (the stationary "solid phase'/"solid substrate" here as opposed to solid microparticle/beads that can be washed away), which is usually constructed as a multiple-well plate known as the "ELISA plate". Conventionally, like other forms of immunoassays, the specificity of antigen-antibody type reaction is used because it is easy to raise an antibody specifically against an antigen in bulk as a reagent. Alternatively, if the analyte itself is an antibody, its target antigen can be used as the binding reagent (Wikipedia article: ELISA).

# 8. Protocols

## 8.1. Immunofluorescence staining/Antibody labelling

| Reagent | Time |
| --- | --- |
| wash with 0.1M PB | 3 x 10' |
| wash with 0.1M TBTX | 3 x 10' |
| incubate in primary antibody solution (0,2% anti-CHH T5B4/70 in 0.1M TBTX, containing 0.02% Na-azide) | overnight |
| wash with 0.1M TBTX | 3 x 15' |
| incubate in second antibody solution (1% GAR-FITC in 0.1M TBTX containing 0.02% Na-azide) | 2 h (RT) |
| wash with 0.1M TBTX | 3 x 10' |

## 8.2. Peroxidase-Antiperoxidase

| Reagent | Time |
| --- | --- |
| rinse in 0.1M TBTX | 3 x 10' |
| preincubate in 2% NGS in 0.1M TBTX | 1 h (RT) |
| incubate in primary antibody solution (0.0125% anti-OCHH T5 B/4 in 0.5M TBTX, containing 1% NGS and 0.02% Na-azide) | overnight (RT) |
| rinse in 0.1M TBTX | 3 x 10' |

| Reagent | Time |
| --- | --- |
| incubate in second antibody solution (1% GAR in 0.1M TBTX, containing 1% NGS and 0.02% Na-azide) | 1 h (RT) |
| rinse in 0.1M TBTX | 3 x 10' |
| incubate in PAP solution (0,33% PAP in 0.5M TBTX containing 1% NGS) | 1 h (RT) |
| rinse in 0.1M TBTX | 3 x 10' |
| rinse in 0.1M PB | 1 x 10' |
| incubate in DAB solution (10mg DAB (first dilute it in 500µl Aqua dest.) in 30ml 0.1M PB with 10µl $H_2O_2$) | 10'-45' (RT) |

## 9. Other

<u>Pipetting pattern for ELISA</u>

|   | 1 | 2 | 3 | 4 | 5 | 6 | 7 | 8 | 9 | 10 | 11 | 12 |
| --- | --- | --- | --- | --- | --- | --- | --- | --- | --- | --- | --- | --- |
| A | Bl | Bl | 10 | 10 | 18 | 18 | 26 | 26 | 34 | 34 | 42 | 42 |
| B | 3 | 3 | 11 | 11 | 19 | 19 | 27 | 27 | 35 | 35 | 43 | 43 |
| C | 4 | 4 | 12 | 12 | 20 | 20 | 28 | 28 | 36 | 36 | 44 | 44 |
| D | 5 | 5 | 13 | 13 | 21 | 21 | 29 | 29 | 37 | 37 | 45 | 45 |
| E | 6 | 6 | 14 | 14 | 22 | 22 | 30 | 30 | 38 | 38 | 46 | 46 |
| F | 7 | 7 | 15 | 15 | 23 | 23 | 31 | 31 | 39 | 39 | 47 | 47 |
| G | 8 | 8 | 16 | 16 | 24 | 24 | 32 | 32 | 40 | 40 | 48 | 48 |
| H | 9 | 9 | 17 | 17 | 25 | 25 | 33 | 33 | 41 | 41 | 49 | 49 |

BI = blank (10µl aceto nitrile)

3 - 49 = sample (10 µl of each sample); for example: 3 = 10µl of sample number 3

## 10. Further Results

<u>Direct ELISA</u>

Direct ELISA 405nm  03.10.12;  10 LemCCCA from 27.09.2012;
2NHac extract HPLC uBondPh    30-80%B60   antiOCHH 1:8k
1. Measurement:   after 15 minutes at 16:59:32

Original Data

|   | 1 | 2 | 3 | 4 | 5 | 6 | 7 | 8 | 9 | 10 | 11 | 12 |
|---|---|---|---|---|---|---|---|---|---|----|----|----|
| A | 0,180 | 0,069 | 0,086 | 0,114 | 0,073 | 0,170 | 0,240 | 0,069 | 0,107 | 0,071 | 0,068 | 0,071 |
| B | 0,072 | 0,070 | 0,069 | 0,179 | 0,382 | 0,070 | 0,077 | 0,083 | 0,068 | 0,069 | 0,136 | 0,068 |
| C | 0,064 | 0,137 | 0,064 | 0,359 | 0,062 | 0,060 | 0,067 | 0,279 | 1,093 | 0,908 | 0,062 | 0,064 |
| D | 0,074 | 0,072 | 0,067 | 0,066 | 0,067 | 0,069 | 0,075 | 0,075 | 0,094 | 0,094 | 0,068 | 0,070 |
| E | 0,073 | 0,154 | 0,071 | 0,371 | 0,070 | 0,072 | 0,077 | 0,074 | 0,073 | 0,074 | 0,068 | 0,094 |
| F | 0,066 | 0,290 | 0,062 | 0,140 | 0,067 | 0,182 | 0,564 | 0,389 | 0,067 | 0,067 | 0,069 | 0,069 |
| G | 0,070 | 0,070 | 0,069 | 0,066 | 0,071 | 0,069 | 0,071 | 0,071 | 0,069 | 0,066 | 0,069 | 0,129 |
| H | 0,070 | 0,070 | 0,067 | 0,160 | 0,074 | 0,072 | 0,070 | 0,071 | 0,067 | 0,082 | 0,063 | 0,082 |

Arithmetic average and blank adjusted

|   | 1 | 2 | 3 | 4 | 5 | 6 | 7 | 8 | 9 | 10 | 11 | 12 |
|---|---|---|---|---|---|---|---|---|---|----|----|----|
| A | 0,000 | 0,000 | -0,025 | -0,025 | -0,003 | -0,003 | 0,030 | 0,030 | -0,036 | -0,036 | -0,055 | -0,055 |
| B | -0,054 | -0,054 | -0,001 | -0,001 | 0,101 | 0,101 | -0,045 | -0,045 | -0,056 | -0,056 | -0,023 | -0,023 |
| C | -0,024 | -0,024 | 0,087 | 0,087 | -0,064 | -0,064 | 0,049 | 0,049 | 0,876 | 0,876 | -0,062 | -0,062 |
| D | -0,052 | -0,052 | -0,058 | -0,058 | -0,057 | -0,057 | -0,050 | -0,050 | -0,031 | -0,031 | -0,056 | -0,056 |
| E | -0,011 | -0,011 | 0,096 | 0,096 | -0,054 | -0,054 | -0,049 | -0,049 | -0,051 | -0,051 | -0,044 | -0,044 |
| F | 0,053 | 0,053 | -0,024 | -0,024 | 0,000 | 0,000 | 0,352 | 0,352 | -0,058 | -0,058 | -0,056 | -0,056 |
| G | -0,055 | -0,055 | -0,057 | -0,057 | -0,055 | -0,055 | -0,054 | -0,054 | -0,057 | -0,057 | -0,026 | -0,026 |

| H | -0,055 | -0,055 | -0,011 | -0,011 | -0,052 | -0,052 | -0,054 | -0,054 | -0,050 | -0,050 | -0,052 | -0,052 |

Direct ELISA 405nm  03.10.12;  10 LemCCCA from 27.09.2012;
2NHac extract HPLC uBondPh    30-80%B60   antiOCHH 1:8k
2. Measurement:  after 30 minutes at 17:14:44

Original Data

| | 1 | 2 | 3 | 4 | 5 | 6 | 7 | 8 | 9 | 10 | 11 | 12 |
|---|---|---|---|---|---|---|---|---|---|---|---|---|
| A | 0,084 | 0,081 | 0,081 | 0,075 | 0,082 | 0,175 | 0,081 | 0,081 | 0,123 | 0,084 | 0,080 | 0,078 |
| B | 0,081 | 0,079 | 0,075 | 0,189 | 0,084 | 0,078 | 0,082 | 0,091 | 0,078 | 0,078 | 0,160 | 0,078 |
| C | 0,071 | 0,157 | 0,066 | 0,260 | 0,066 | 0,065 | 0,084 | 0,084 | 1,750 | 1,577 | 0,068 | 0,068 |
| D | 0,080 | 0,081 | 0,074 | 0,072 | 0,074 | 0,073 | 0,084 | 0,079 | 0,121 | 0,123 | 0,074 | 0,078 |
| E | 0,087 | 0,089 | 0,079 | 0,074 | 0,082 | 0,078 | 0,090 | 0,087 | 0,085 | 0,087 | 0,077 | 0,094 |
| F | 0,076 | 0,071 | 0,068 | 0,071 | 0,074 | 0,187 | 0,647 | 0,590 | 0,074 | 0,075 | 0,080 | 0,082 |
| G | 0,078 | 0,077 | 0,075 | 0,074 | 0,079 | 0,074 | 0,080 | 0,082 | 0,076 | 0,077 | 0,078 | 0,079 |
| H | 0,076 | 0,078 | 0,074 | 0,098 | 0,082 | 0,079 | 0,081 | 0,082 | 0,075 | 0,074 | 0,067 | 0,077 |

Arithmetic average and blank adjusted

| | 1 | 2 | 3 | 4 | 5 | 6 | 7 | 8 | 9 | 10 | 11 | 12 |
|---|---|---|---|---|---|---|---|---|---|---|---|---|
| A | 0,000 | 0,000 | -0,004 | -0,004 | 0,046 | 0,046 | -0,001 | -0,001 | 0,021 | 0,021 | -0,003 | -0,003 |
| B | -0,002 | -0,002 | 0,050 | 0,050 | -0,001 | -0,001 | 0,004 | 0,004 | -0,004 | -0,004 | 0,037 | 0,037 |
| C | 0,032 | 0,032 | 0,080 | 0,080 | -0,017 | -0,017 | 0,002 | 0,002 | 1,581 | 1,581 | -0,014 | -0,014 |
| D | -0,002 | -0,002 | -0,009 | -0,009 | -0,009 | -0,009 | -0,001 | -0,001 | 0,040 | 0,040 | -0,006 | -0,006 |
| E | 0,006 | 0,006 | -0,006 | -0,006 | -0,002 | -0,002 | 0,006 | 0,006 | 0,003 | 0,003 | 0,003 | 0,003 |
| F | -0,009 | -0,009 | -0,013 | -0,013 | 0,048 | 0,048 | 0,536 | 0,536 | -0,008 | -0,008 | -0,001 | -0,001 |
| G | -0,005 | -0,005 | -0,008 | -0,008 | -0,006 | -0,006 | -0,001 | -0,001 | -0,006 | -0,006 | -0,004 | -0,004 |
| H | -0,005 | -0,005 | 0,003 | 0,003 | -0,002 | -0,002 | -0,001 | -0,001 | -0,008 | -0,008 | -0,010 | -0,010 |

Direct ELISA 405nm  03.10.12;  10 LemCCCA from 27.09.2012;
2NHac extract HPLC uBondPh    30-80%B60   antiOCHH 1:8k
3. Measurement: after 45 minutes at 17:35:53

Original Data

| | 1 | 2 | 3 | 4 | 5 | 6 | 7 | 8 | 9 | 10 | 11 | 12 |
|---|---|---|---|---|---|---|---|---|---|---|---|---|
| A | 0,100 | 0,094 | 0,107 | 0,088 | 0,093 | 0,194 | 0,094 | 0,094 | 0,144 | 0,100 | 0,090 | 0,090 |
| B | 0,093 | 0,088 | 0,084 | 0,085 | 0,094 | 0,091 | 0,095 | 0,090 | 0,087 | 0,088 | 0,083 | 0,088 |
| C | 0,081 | 0,077 | 0,075 | 0,075 | 0,077 | 0,073 | 0,100 | 0,104 | 2,623 | 2,319 | 0,078 | 0,078 |

| | 1 | 2 | 3 | 4 | 5 | 6 | 7 | 8 | 9 | 10 | 11 | 12 |
|---|---|---|---|---|---|---|---|---|---|---|---|---|
| D | 0,088 | 0,097 | 0,084 | 0,081 | 0,081 | 0,084 | 0,099 | 0,092 | 0,157 | 0,159 | 0,082 | 0,091 |
| E | 0,104 | 0,101 | 0,092 | 0,089 | 0,095 | 0,093 | 0,111 | 0,105 | 0,104 | 0,099 | 0,086 | 0,103 |
| F | 0,086 | 0,080 | 0,079 | 0,078 | 0,086 | 0,205 | 0,969 | 0,904 | 0,085 | 0,084 | 0,083 | 0,087 |
| G | 0,093 | 0,087 | 0,086 | 0,080 | 0,088 | 0,085 | 0,095 | 0,097 | 0,087 | 0,084 | 0,084 | 0,086 |
| H | 0,086 | 0,089 | 0,080 | 0,110 | 0,092 | 0,090 | 0,095 | 0,093 | 0,085 | 0,083 | 0,074 | 0,086 |

Arithmetic average and blank adjusted

| | 1 | 2 | 3 | 4 | 5 | 6 | 7 | 8 | 9 | 10 | 11 | 12 |
|---|---|---|---|---|---|---|---|---|---|---|---|---|
| A | 0,000 | 0,000 | 0,000 | 0,000 | 0,046 | 0,046 | -0,003 | -0,003 | 0,025 | 0,025 | -0,007 | -0,007 |
| B | -0,007 | -0,007 | -0,013 | -0,013 | -0,005 | -0,005 | -0,005 | -0,005 | -0,010 | -0,010 | -0,012 | -0,012 |
| C | -0,018 | -0,018 | -0,022 | -0,022 | -0,022 | -0,022 | 0,005 | 0,005 | 2,374 | 2,374 | -0,019 | -0,019 |
| D | -0,005 | -0,005 | -0,015 | -0,015 | -0,015 | -0,015 | -0,002 | -0,002 | 0,061 | 0,061 | -0,010 | -0,010 |
| E | 0,006 | 0,006 | -0,007 | -0,007 | -0,003 | -0,003 | 0,011 | 0,011 | 0,005 | 0,005 | -0,002 | -0,002 |
| F | -0,014 | -0,014 | -0,019 | -0,019 | 0,049 | 0,049 | 0,839 | 0,839 | -0,013 | -0,013 | -0,012 | -0,012 |
| G | -0,007 | -0,007 | -0,014 | -0,014 | -0,010 | -0,010 | -0,001 | -0,001 | -0,012 | -0,012 | -0,012 | -0,012 |
| H | -0,009 | -0,009 | -0,002 | -0,002 | -0,006 | -0,006 | -0,003 | -0,003 | -0,013 | -0,013 | -0,017 | -0,017 |

## 11. References

Bovine OPG 2008. Datasheet BO0027MSDS.

Dircksen H 2008. Review. Insect ion transport peptides are derived from alternatively spliced genes and differentially expressed in the central and peripheral nervous system.

Dircksen H, Tesfai LK, Albus C, Nässel DR 2008. Ion Transport Peptide Splice Forms in Central and Peripheral Neurons Throughout Postembryogenesis of Drosophila melanogaster.

Encyclopaedia Britannica article: Neurohormone. http://www.britannica.com/EBchecked/topic/410635/neurohormone 03.10.12 17:51.

Haugland RP 2002. Handbook of Fluorescent Probes and Research Products, 9th edition. Molecular Probes, p.1.

Junqueira LC, Carneiro J 2003. Basic Histology, Text & Atlas, 10th edition. Lange, p.9-10.

Lorenzon S 2005. Hyperglycemic stress response in Crustacea.

Lucas J 2006. Comparative investigations on the localization and identification of ion transport peptide (ITP) of the cockroach and the fruit fly.

Sternberger LA 1974. Immunocytochemistry. p.110-111.

Webster SG, Keller R, Dircksen H 2011. The CHH-superfamily of multifunctional peptide hormones controlling crustacean metabolism, osmoregulation, moutling, and reproduction.

Wikipedia article: ELISA. http://en.wikipedia.org/wiki/ELISA 04.10.12 18:09.

Wikipedia article: Fixation (histology). http://en.wikipedia.org/wiki/Fixation_(histology) 03.10.21 17:45.

Wikipedia article: HPLC. http://en.wikipedia.org/wiki/High-performance_liquid_chromatography 04.10.12 17:58.

Wikipedia article: Immunohistochemistry. http://en.wikipedia.org/wiki/Immunohistochemistry 04.10.12 16:11.

Wikipedia article: Mushroom bodies. http://en.wikipedia.org/wiki/Mushroom_bodies 20.10.12 17:12